IMAGE EVALUATION
TEST TARGET (MT-3)

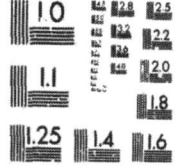

← 6″ →

Photographic
Sciences
Corporation

23 WEST MAIN STREET
WEBSTER, N.Y. 14580
(716) 872-4503

CIHM/ICMH Microfiche Series.

CIHM/ICMH Collection de microfiches.

Canadian Institute for Historical Microreproductions / Institut canadien de microreproductions historiques

Technical and Bibliographic Notes/Notes techniques et bibliographiques

The Institute has attempted to obtain the best original copy available for filming. Features of this copy which may be bibliographically unique, which may alter any of the images in the reproduction, or which may significantly change the usual method of filming, are checked below.

L'Institut a microfilmé le meilleur exemplaire qu'il lui a été possible de se procurer. Les détails de cet exemplaire qui sont peut-être uniques du point de vue bibliographique, qui peuvent modifier une image reproduite, ou qui peuvent exiger une modification dans la méthode normale de filmage sont indiqués ci-dessous.

- [✓] Coloured covers/
  Couverture de couleur

- [ ] Covers damaged/
  Couverture endommagée

- [ ] Covers restored and/or laminated/
  Couverture restaurée et/ou pelliculée

- [ ] Cover title missing/
  Le titre de couverture manque

- [ ] Coloured maps/
  Cartes géographiques en couleur

- [ ] Coloured ink (i.e. other than blue or black)/
  Encre de couleur (i.e. autre que bleue ou noire)

- [ ] Coloured plates and/or illustrations/
  Planches et/ou illustrations en couleur

- [ ] Bound with other material/
  Relié avec d'autres documents

- [ ] Tight binding may cause shadows or distortion along interior margin/
  La reliure serrée peut causer de l'ombre ou de la distortion le long de la marge intérieure

- [ ] Blank leaves added during restoration may appear within the text. Whenever possible, these have been omitted from filming/
  Il se peut que certaines pages blanches ajoutées lors d'une restauration apparaissent dans le texte, mais, lorsque cela était possible, ces pages n'ont pas été filmées.

- [ ] Additional comments:/
  Commentaires supplémentaires:

- [ ] Coloured pages/
  Pages de couleur

- [ ] Pages damaged/
  Pages endommagées

- [ ] Pages restored and/or laminated/
  Pages restaurées et/ou pelliculées

- [ ] Pages discoloured, stained or foxed/
  Pages décolorées, tachetées ou piquées

- [ ] Pages detached/
  Pages détachées

- [ ] Showthrough/
  Transparence

- [✓] Quality of print varies/
  Qualité inégale de l'impression

- [ ] Includes supplementary material/
  Comprend du matériel supplémentaire

- [ ] Only edition available/
  Seule édition disponible

- [ ] Pages wholly or partially obscured by errata slips, tissues, etc., have been refilmed to ensure the best possible image/
  Les pages totalement ou partiellement obscurcies par un feuillet d'errata, une pelure, etc., ont été filmées à nouveau de façon à obtenir la meilleure image possible.

This item is filmed at the reduction ratio checked below/
Ce document est filmé au taux de réduction indiqué ci-dessous.

10X		14X		18X		22X		26X		30X	
						✓					
	12X		16X		20X		24X		28X		32X

1	2	3

1
2
3

1	2	3
4	5	6

THE
Flora of Newfoundland, Labrador and St. Pierre et Miquelon,
PART III.

By THE REV. ARTHUR C. WAGHORNE,

BAY OF ISLANDS, NEWFOUNDLAND.

From the Transactions of the Nova Scotian Institute of Science,
Vol. IX, Session 1897-98.

Sold by S. E. Garland
New & Second Hand Books
St. Johns, NFLD.

Price 20 Cents.

VIII.—THE FLORA OF NEWFOUNDLAND, LABRADOR, and ST. PIERRE ET MIQUELON: PART III.—BY THE REV. ARTHUR C. WAGHORNE, *Bay of Islands, Newfoundland.*

(Read May 9th, 1898.)

A change of residence, and the charge of a new and extensive parish, are the hindrances which are chiefly accountable for the delay in the continuance of this series of papers, of which the first part appeared in the *Transactions* of the Institute Vol. VIII (Ser. 2, Vol. I), page 359; and the second part in Vol. IX (Ser. 2, Vol. II.), page 17.

To the prefatory remarks of the two preceding papers, a few notes should here be added :—

1. The Part I. here first presented is made up of Polypetalous plants which have been added to our Newfoundland or Labrador lists of plants since 1895.

2. These have been obtained from three sources, chiefly :—

(*a*) Professor Macoun's *Contributions from the Herbarium of the Geological Survey of Canada*, (Parts I.—VI.); these are indicated herein briefly as " C. H. Geo. S. of C."

(*b*) A list of Newfoundland plants collected by Mr. A. B. Bullman, B. A. Sc. (of H. M. Newfoundland Survey). The plants were collected by him on the West coast in 1896, and in White Bay in 1897. This gentlemen modestly says that he lays no claim to be a botanist, so that his determinations may be subject to revision, a fate which befalls even those of men who *are* botanists. If Mr. Bullman's decisions are sustained as to certain plants, his list adds fifteen names to our flora.

(*c*) My own collections since 1895 in the Bay of Islands, and a week while in Bay St. George, and a trip across the country in 1895, and to Brenton and Clode Sound in Bonavista Bay, on the East coast. At Mr. Reid's stone quarry, about 80 miles from the Bay of Islands, on both sides of the railway track, by the sides of the ponds in the bogs, I was fortunate enough to find, for the first time, the Schizæa pusilla, Pursh, in fair quantity.

3. To the gentlemen who have kindly assisted me in the determination of my plants, I may now add the names of Dr. Wm. Trelease of St. Louis, Dr. B. L. Robinson of Harvard, and Mr. T. V. Coville of Washington.

4. If permitted so to do, I would gladly say that I have been for the last three or four years distributing my plants, including mosses and lichens, (the fungi I hope to have ready this coming season), and that I should be glad to hear of any who may desire specimens.

I.—POLYPETALÆ. (Supplementary to Parts I. and II.)

I.—RANUNCULACEÆ.

265. *Anemone nemorosa*, L. WOOD ANEMONE. Deer Arm, Bonne Bay (Bulhman). Moist places. July.

7. *Ranunculus abortivus*, L. Var. *micranthus*, Gray. Chimney Cove, B. of I., (A. C. W.—Fowler). Sea cliffs. June.

266. *R. septentrionalis*, Poir. Near Meadows, B. of I. (A. W. Trelease), and Deer Lake (according to Dr. Robinson, but this plant was named R. Macounii by Dr. Trelease). Meadows and wet places. July.

267. *R. Macounii*, Brit. Chimney Cove, B. of I. (A. C. W.—Trelease). Fields. July.

268. *R. fascicularis*, Muhl. Birchy Cove, B. of I. (A. C. W.—Trelease. Fields. June. (Dr. Robinson says this is R. repens, L.)

19. *R. recurvatus*, Poir. Humber River (A. C. W.—Fowler). River side. June.

IV.—MENISPERMACEÆ.

269. *Menispermum Canadense*, Pointe Lafontane, West Coast (Bulhman). Low ground. June.

VII.—CRUCIFERÆ.

270. *Erysimum asperum*, D. C. Chimney Cove, B. of I. (A. C. W.—Fowler). Sea cliffs. June.

271. *Arabis lævigata*, Poir. SMOOTH ROCK CRESS. Cow Head (Bullman). Dry, rocky ground. June.

36. *Barbarea vulgaris*, R. Br. Mentioned in Prof. Macoun's C. H. Geo. S. of C. (V. 2) as not before recorded from the Labrador, was twice collected by the compiler in 1894 in the Straits of Belle Isle. *See* Part II., p. 29.

138. *Nasturtium palustre* D. C. To the Labrador reference in Part I., p. 29, may now be added its Newfoundland occurrence at Deer Lake (A. C. W.—Fowler). River side. August.

272. *Cardamine Pennsylvanica*, Muhl. Whitbourne, appearing introduced (R. & S.).

273. *Hesperis matronalis*, L. Street of St. John's; infrequent (R. & S.).

274. *Subularia aquatica*, L. Terrestrial form. Exploits River and around pond near Whitbourne (R. & S.). August.

VIII.—VIOLACEÆ.

62. *Viola palustris*, L. Though recorded in Part I., p. 9, from the Labrador (named by himself), this is mentioned in Prof. Macoun's C. H. Geo. S. of C. (V. 2), as not before (1894) recorded from Labrador.

65. *V. canina*, L. *Muhlenbergii*. In Bay of Islands at Birchy Cove (Fowler), Coal River and Lark Harbour (Robinson), and Rope Cove (Trelease). Fields and wet places. June.

XIV.—CARYOPHYLLACEÆ.

275. *Silene Armeria*, L. Deer Lake (A. C. W.—Fowler) Near gardens. August.

276. *Arenaria Sajanensis*, Willd. Coal River, near B. of I. (A. C. W.—Robinson). Sandy places. July. *Lab*: Cape Chidley (C. H. Geo. S. of C., V. 5).

69. *A. verna*, L. Englee, White Bay, and Coal River, B. of I. (A. C. W.—Fowler). June.

var. hirta, Bigel. Coal River (A. C. W.—Fowler). Sandy places. June.

277. *A. patula*, Mx. Chimney Cove, B. of I. Hills. June. (Dr. Robinson says, *A. verna*, L.).

278. *A. ciliata*, L., var. *A. humifusa*, Hornem. Coal River (A. C. W.—Robinson). Sandy places. July.

89. *Stellaria borealis*, Bigel. Var. *alpestris*, Gray. Shoal Point, B. of I. (A. C. W.—Trelease); Dr. Robinson says " this is very different from Dr. Gray's conception and is much better considered only as *S. borealis*, L."

XVI.—HYPERICACEÆ.

104. *Hypericum Canadense*, L. Var. *miserrimum*, Chois Seal Rocks, Sandy Point, Bay St. George (A. C. W.—Fowler) Wet places. August.

XVII.—GERANIACEÆ.

279. *Geranium Robertianum*, L. Chimney Cove, near Bay of I. (A. C. W.—Fowler). Ravines. July.

110. *Oxalis acetosella*, L. To the doubtful occurrence referred to in my Part I, 13, may now be added Portland Creek, White Bay (Bullman). Woods. June.

XX.—LEGUMINOSÆ.

280. *Astragalus astragalinus*, D. C. *Minnesota Botanical Studies*. Bull. IX., p. 65.

117. *A. alpinus*. L. Chimney Cove, B. of I. (A. C. W.—Trelease). High seacliffs. July. Professor Shelden (St. Louis) refers this to *Spiesia Lamberti* (Pursh), *v. spicata*. Hook.

124. *Oxytropis campestris*, D. C. Chimney Cove (A. C. W. Fowler and Robinson). Hills. July and August. And White Bay (Bullman). Stony Shore.

119. *Hedysarum boreale*, Muhl. Wild Cove, near B. of I. (A. C. W.—Trelease). Sea bank. July.

XXXII.—Rosaceæ.

281. *Gillenia trifoliata*, Moench. Ponds River, West Coast. (Bullman). Woods. June.

282. *Agrimonia striata*, Mx. Chimney Cove, B. of I. (A. C. W.—Shedden). Hills. July. Reported as *A. Eupatoria* by Prof. Fowler.

283. *A. hirsuta*, Wilds. Chimney Cove (A. C. W.—Robinson). High hills. August.

253. *Fragaria vesca*, L, var. *Americana*, Porter. Chimney Cove (A. C. W.—Trelease). Banks. July.

284. *Geum album*, Gmel. Cow Head, Bonne Bay (Bullman). Woods. June.

285. *Potentilla littoralis*, Rydberg. Chimney Cove, B. of I. (A. C. W.—Robinson). Hills. August. Reported by Dr. Fowler as *P. Pennsylvanica*, L.

179. *Rosa Carolina*, L. The reference to this from New Harbor (Part. II., p. 25) should have been marked as very doubtfully named by Prof. Macoun. M. Crepin says it must be omitted.

286. *R. humilis*, Marsh. Shoal Point, near Bay of Islands (A. C. W.—Dr. G. N. Bert). St. John's (Robinson and Schenk). Woods. August.

XXXIII.—Saxifragaceæ.

201. *Ribes rubrum*, L, var. *subglandulosum*, Maxim. Crabbes (Fowler) and Frenchman's Cove (Robinson), B. of I. (A. C. W.) Woods. June.

287. *Ribes floridum*, *Wild Black Currant*. Cow Head (Bullman). June.

197. *Parnassia parviflora*, D. C. Chimney Cove (Robinson), and Goose Arm, B. of I. (Fowler). A. C. W. Hills and wet banks. August.

XXXV.—Droseraceæ.

220. *Drosera intermedia*, Drev. and Hayne. Var. *Americana*, D. C. *Lab:* Upper West Branch, Hamilton River (A. P. Low, 1894, C. H. Geo. S. of C., V. 6.)

XXVIII.—Rhamnaceæ *(Blackthorn Family).*

288. *Rhamnus alnifolia,* L'Her.   Deer Lake (Fowler) and Lark Harbour, B. of I. (A. C. W.).   Woods.   July.

XXXVI.—Hamamelaceæ *Water-hazel Family*

289. *Hamamelis Virginiana,* L.   Cow Head, Bonne Bay (Bullman).   Woods.   July.

XL.—Lythraceæ.

290. *Nesæa verticillata,* H. B. K. ?   St. John's Island, St. John's Bay, West Coast.   (Bullman).   July.

XLV.—Umbelliferæ.

250. *Osmorrhiza brevistylis,* D. C.   Frenchman's Cove, B. of I. (A. C. W.—Fowler).   Woods.   June and July.

251. *Sanicula Canadensis,* L.   Wild Cove (Trelease), and Shoal Point, B. of I. (Fowler) A. C. W.   Banks.   July.   Mr. Fernand says of the Wild Cove specimen that it is better considered as *S. Marylandica,* L.

GAMOPETALÆ.

XLVIII.—Caprifoliaceæ *Honeysuckle Family.*

275. *Diervilla trifida,* Mœnch.   Bush Honeysuckle   S. John's.   Common (Robinson and Schreuk).   Brigus (Bell, Cat. III., 540); Nipper's Harbour (Notre Dame Bay) and near Badger's Brook, Exploits River (both of these collected by Revd. J. H. Bull, named by Messrs. Macoun and Fowler)   Topsail (A. C. W.—Macoun).   Rocky hills.   July and August.   *Flora Miq.,* on little hillocks near the Brook Sylvain

276. *Linnæa borealis,* L.   Twin-flower (GROUND-IVY AND TRUMPET-FLOWER) appear to be pretty common throughout Newfoundland and the Labrador, in mossy woods.   Reported from Trinity Bay, S. John's, Fortune Bay White Bay, Bay St. George, and Bay of Islands.   *Lab.*: Hopedale (Weiz-Packard);

collected at Pack's Harbour and Fortean by myself and by Mr.
Butler, on hillside. July and August. *Flora Miq.*

277. *Lonicera cerulea* L. *Mountain Honeysuckle* (Cat. II.,
198); (Rocks), West of Random. In full fruit September 10,
Cormack, S. John's (Miss Southcott); New Harbour and Green's
Harbour, (Trinity Bay), and Harbour Breton (Fortune Bay)
collected by myself and named by Prof. Macoun. Rope Cove,
B. of I. (Fowler), Whitbourne Robinson and Schrenk; Long
Point, north of Cape St. George (Bell). Bogs and wet woods.
July and August. *Lab*; (Cat. II., 198), found by myself at
Indian Harbour (near Battle Harbour), and Battle Harbour, and
L'anse au Loup, and Blanc Sablon, determined by Professors
Macoun and Fowler. Caribou Island (Butler-Packard). July.

Var. villosa, T. & G., (Cat. II., 198.) Newfoundland
(Pylaie), Coast of *Labrador* (McGill College, Herb.).

278. *L. oblongifolia*, Hook. *Swamp Honeysuckle.* Great
Cod Roy River (Bell).

279. *Sambucus racemosa*, L. *Red-berried Elder.* Random
(Trinity Bay), Bay Despair (Hermitage Bay), Swan Island
(Notre Dame Bay), and Bay of Islands. Collected by myself
and named by Messrs Macoun, Fowler and Morton. Woods.
July.

Var. pubens, Wats, (which Prof. Macoun [Cat. II., 537]
unites with the last), near Flat Bay Brook (Bell) and S. John's
(Miss Southcott—Macoun).

280. *Viburnum Lentago*, L. *Sweet Viburnum, Sheep
Berry.* Near Flat Bay Brook, Bay St. George, and Humber
River, B. of I. (Bell). June and July.

281. *V. cassinoides*, T. & G. *Withe Rod.* (Cat. II., 194).
This appears to be common in woods throughout the country. I
have it reported from Trinity Bay, S. John's, Bay Despair, Hermitage Bay, and several places about Bay of Islands. July and
August. *Flora Miq.*

282. *V. pauciflorum*, Pylaie. *Few-Flowered Viburnum or
Arrow-wood* (Squash Berry). Several places in Fortune and

Trinity Bays, Bay of Islands, S. John's (Miss Southcott). Petty Harbour (Reeks). Great Cod Roy River, Bay S. George (Bell). Manuel's River, Conception Bay (Robinson and Schrenk). Rocky banks near brooks in woods. June—August. *Lab:* ravines (Butler, Cat. II., 195.)

283. *V. Opulus*, L. *High-Bush Cranberry* (called *TRASH BERRY, I believe by Newfoundlanders mostly). Not so common as the last. Topsail (Bell, Cat. III., 539). Swamps near confluence of Exploit and Badger's River (Robinson and Schrenk). Great Cod Roy River (Bay S. George). Humber River and island in Deer Pond (B. of I.). Bell; (Reeks). July and August.

284. *V. acerifolium*, L. *Maple-leaved Arrow-wood.* Near Flat Bay Brook, Bay St. George (Bell). June.

XI.—RUBIACEÆ. *Madder Family.*

285. *Galium boreale*, L. *Northern Bedstraw* (Reeks).

286. *G. asprellum*, Mx. *Rough Bedstraw.* Whitbourne (Robinson and Schrenk); Bay Bull's Arm (Trinity Bay) and Shoal Point, B. of I. (A. C. W.—Macoun and Fowler); (Reeks). Woods and moist ground. August.

287. *G. trifidum*, Mx. *Small Bedstraw.* In the Bay of Islands, collected by myself, at Riverhead (Fowler), and Lark Harbour (Robinson). Benton, Bonavista Bay (Fowler); several places in Trinity Bay, and Topsail in Conception Bay (Fowler and Macoun). Wayside and wet places. August. *Lab:* Hopedale (Weiz-Packard); Forteau (A. C. W.—Eaton). August.

Var. pusillum, Gray. Sunny banks, Whitbourne (Robinson and Schrenk); White Bay or Ferryland (Revd. R. Temple—Fowler). *Lab:* Forteau (A. C. W.—Fowler and Butler). June and July.

Var. tinctorium, T. & G. (Cat. II., 201).

Var. latifolium, Torr. Open woods, S. John's (Robinson and Schrenk). August.

---

* The "Joint-wood Berry" and "White-wood Berry" of Newfoundland, must, I think, be this, or one other of this genus.

288. *G. Aparine*, L. *Goosegrass.* Middle Arm, B. of I. (A. C. W.—Fowler). Woods. July.

289. *G. triflorum*, Mx. *Three-flowered Galium.* Near confluence of Exploits and Badger's River (Robinson and Schrenk); New Harbour, Trinity Bay (Macoun); and Wild Cove (Fowler), and Chimney Cove (Trelease). Woods. July. *Lab:* Capstan Island (A. C. W.—Eaton). Open woods. July.

290. *G. Mollugo*, L. *Narrow-leaved Bedstraw.* S. John's (Robinson and Schrenk). Ploughed ground. August.

291. *G. palustre*, L. var. *minus*, Lge. *Lab:* Long Point, Hamilton Island (A. C. W.—Macoun), (C. H. Geo. S. of C., IV. 202).

292. *Mitchella repens*, L. *Partridge berry* (Reeks); Flat Bay Brook (Bell). *Flora Miq.*, found once near stream Bibite, in damp ground, in August. Sought after by the partridges of the island.

I.—VALERIANACEÆ.   *Valerian Family.*

293. *Valeriana dioica*, L. var. *sylvatica*, Wats. *Marsh Valerian* (Banks, Cat. III., 204).

III.—COMPOSITÆ.   *Composite Family.*

294. *Achillea Millefolium*, L. *Yarrow. Milfoil (Dead Man's Daisy).* Common in fields. *Lab:* Forteau. August, (rose-coloured). *Flora Miq.*, common, July, August.

Var. nigrescens, E. Meyer. *Lab:* Nain (Bell, Cat. III., 552), Caribou Island (Butler), and Hopedale (Weiz-Packard.)

295. *A. Ptarmica*, L. Introduced, Harbour Grace (McGill Coll. Herb.—Cat. II., 251).

296. *Arnica alpina*, Murr. *Lab:* Nachvak and Cape Chimney (Bell) Cat II., 261; III. 535.

297. *Antennaria alpina*, L. *Everlasting. Lab:* (Kohmeister —Cat. II., 236), Hopedale (Weiz), and Caribou Island (Butler) Packard.

298. *A. Carpathica*, R. Br. Wet, boggy places and river margins, (Gray) Cat. II., 237.

299. *A. dioica*, Gaertn. *Mountain Cudweed or Everlasting*. From Newfoundland to *Labrador*, and the extreme Arctic regions and dry mountain pastures of the Rocky Mountains (Cat. II., 236). S. John's (Miss Southcott); Exploits (A. C. W. —Fowler). Rocky banks. July.

300. *A. plantaginifolia*, Hook. (Reeks), Badger's Brook (Revd. I. H. Bull —Fowler); Middle Arm, B. of I. Rocky bed of S. W. Arm River. Holyrood (Robinson and Schrenk). Sea cliff. Open Woods. June to September.

301. *Anaphalis margaritacea*, Benth and Hook. *Pearly Everlasting.* (Reeks). Common on dry soil along the margins of fields and borders of woods, from Newfoundland to the Pacific (Cat. II., 237). Rocky hills, S. John's and Holyrood (Robinson and Schrenk); Bay S. George (Howley and Bell). Common about B. of I. and Trinity Bay (A. C. W.) July.

302. *Artemisia borealis*, Pall. var. *spithamaea*, T. & G. *Lab.*; (Kohneister, Cat. II., 255). Hopedale Island (Weiz-Packard).

303. *A. Canadensis*, Mx. Near Badger's Brook (Revd. I. H. Bull—Fowler); in the Bay of Islands, Middle Arm, near Rope Cove, Chimney Cove and Goose Arm (A. C. W.—Fowler and Robinson). Woods, sea cliffs, and sandy plains; dry river bed. July and August.

304. *A. Absinthium*, L. *Wormwood.* Naturalized in numerous places by roadsides, in lanes, and about dwellings, from Newfoundland to the western part of Ontario. (Gray, Cat. II., 259).

305. *Arctium Lappa*, L. *Burdock.* S. John's (Professor Holloway—Fletcher). In leaf only, but probably var. *minus*, Holyrood (Robinson and Schrenk).

306. *Anthemis arvensis*, L. *Wild Chamomile.* Random, Trinity Bay (A. C. W.—Macoun).

307. *A. Cotula*, D. C. Clode's Sound, Bonavista Bay (Robinson and Schrenk) Frenchman's Cove, B. of I. Garden weed and about buildings. August and September.

308. *Aster macrophyllus*, L. *Large-leaved Aster.* I cannot find my authority for this plant.

309. *A. Radula*, Ait. *Rasp-leaved Aster.* (Cat. II., 219). West of Random (Cormack); S. John's (Miss Southcott); New Harbour and S. Anthony, N. E. coast, (collected by myself and named by Prof. Macoun); Benton, Bonavista Bay (A. C. W.—Fowler). Rocky bank Manuel's River (Robinson and Schrenk). *Lab:* (Butler, Cat. II., 219; collected by myself at Battle Harbour and L'anse au Loup (Macoun). August and September. Wet places. *Flora Miq.*, damp marl, rarely dry places; very common.

Var. strictus, Gray. Rocky hills, S. John's; moist open ground at confluence of Exploit and Badger's River; Grand Lake, B. of I. (Robinson). *Lab:* (Pursh, Cat. II., 219); Hopedale (Weiz-Packard); Square Islands and Capstan Island (A. C. W.—Macoun and Eaton). Bogs and wet places. August.

310. *A. lævis*, L. Little Harbour, near B of I. (A. C. W.—Trelease—" inflorescence more corymbose and bracts more foliaceous, and not so green-tipped"). September.

311. *A. paniculatus*, L. Harbour Breton, Fortune Bay, (A. C. W.—Macoun).

312. *A. salicifolius*, Ait. Harbour Breton (A. C. W.—Macoun).

313. *A. Novi-Belgii*, L. Exploits River (Robinson and Schrenk); near Frenchman's Cove and Coal River, B. of I. (A. C. W.—Robinson). The latter was reported as *A. tardiflorus* by Prof. Fowler. Wet places. September. Vide *A. longifolius* below, No. 315.

314. *A. tardiflorus*, L. *Long-leaved Aster.* S. John's, Beach and Lark Harbour, B. of I. (A. C. W.—Fowler); New Harbour, Trinity Bay, and Harbour Breton, Fortune Bay (A. C. W.—Macoun). *Lab:* (Gray, Cat. II., 545); Battle Harbour (Bull); Fox Harbour, Hawk's Bay and Cartwright (A. C. W.—Macoun); L'anse au Mort. (A. C. W.—Fowler). August and September.

315. *A. longifolius*, Law. Near meadows, B. of I. (A. C. W.
—Robinson). Reported by Mr. Covelle as *A. Novæ-Angliæ,
var.* banks. September.

316. *A. puniceus*, L. *Red-stalked Aster* or *Starwort*
(Reeks). New Harbour and Harbour Breton (A. C. W.—
Macoun); river banks, Salmonier, common (Robinson and Schrenk);
Benton, Bonavista Bay (A. C. W.—Fowler). Woods. August
and September. *Lab:* Lake Michikamov (A. L. Low—C. H.
Geo. S. of C., VI., 6); L'anse au Loup (A. C. W.—Fowler).
September.

Var. lucidulus, Gray. Was named by Dr. Trelease and
found in the Bay of Islands.

Var. firmus, T. & G. (= *v. lævicaules*, Gray). A doubtful
specimen of Mr. Howley's from Bay S. George (Macoun). *Lab:*
Deep Water Creek (A. C. W.—Macoun; reported as doubtful).

Var. ? Shoal Point, B. of I. (A. C. W.—Trelease). Woods.
August.

317. *A. acuminatus*, Mx. I cannot give the authority for
this.

318. *A. ptarmicoides*, T. & G. S. John's Island, S. John's
Bay. West Coast (Bullman). Stony hills. July.

319. *A. nemoralis*, Ait. *Wood Aster.* Harbour Grace
Cat. III., 227); West of Random (Cormack); S. John's (Prof.)
Holloway—Fowler); Harbour Breton (Macoun), and Seal Rocks,
Sandy Point, Bay S. George (A. C. W.—Fowler); Balley Hally
bog, S. John's (Robinson and Schrenk). Bogs. August. *Flora
Miq.*, very common.

320. *A. linariifolius*, L. *Double-bristled Aster.* (Cormack,
Cat. II., 229).

321. *A. umbellatus*, Mill (Cat. II., 229) S. John's (Miss
Southcott—Fletcher); S. George's Bay (Howley—Macoun);
collected by the compiler at South Side, Harbour Grace, Harbour
Breton and Bay Bull's Arm; named by Prof. Macoun, and at
Clode Sound (Trelease). Common, especially along streams.

Manuel's, Conception Bay (Robinson and Schrenk). Wet places. August and September.

322. *Bidens frondosa*, L. *Common Bigger-tick, Stick Tight.* Near meadows.

323. *Centaurea Cyanus*, L. *Blue-Bottle.* Introduced. (Reek).

324. *C. nigra*, L. *Black Knapweed (Broad Weed and French Clover)* (Reeks). S. John's (Miss Southcott and R. & S.). Apsey Beach (B. of I.). Fields. July and August.

325. *Cichorium Intybus*, L. *Chicory.* S. John's. Infrequent (R. & S.).

326. *Cnicus lanceolatus*, Hoffm. *Common Roadside Thistle.* Manuel's (R. & S.); Norman's Cove, Trinity Bay, Birchy Cove, B. of I. Cleared ground, roadsides. August.

327. *C. muticus*, Pursh. *Glutinous or Swamp Thistle. Horse Tops*, in White Bay (Reeks). Englee and other places in White Bay, and New Harbour, Trinity Bay; S. Paul's Bay (Bullman); Exploit's River (R. & S.). July—September.

328. *C. pumilus*, Torr. *Pasture Thistle.* Flat Bay (Bell).

329. *C. arvensis*, Pursh. *Canada or Field Thistle.* Hopedale, Trinity Bay (Macoun), and Birchy Cove, B. of I. (A. C. W. —Fowler); Bonne Bay (Bullman); Great Cod Roy River (Bell); S. John's (R. & S.). Wayside. July—September.

330. *Chrysanthemum Leucanthemum*, L. *Great White Large Daisy*, (BACHELOR'S BUTTONS), (Reeks); S. John's, common (R. & S.); Birchy Cove, and a few other places about Bay of Islands. Pastures. July and August. *Lab:* Battle Harbour, and by paths over hill near Forteau Lighthouse (A. C. W.— Fowler). August and September.

331. *Erigeron acris*, L. *Fleabane. Lab:* (Torr and Gray, Cat. II., 234).

Var. *Dræbachianus*, Blytt. *Lab:* (Gray, Cat. II., 235); Hopedale (Weiz-Packard).

Var. *debilis*, Gray. *Lab:* Lab. North, and Hudson Bay (Gray, Cat. III., 548).

332. *E. Canadense*, L. *Fireweed* (Rocks).

333. *E. strigosus*, Muhl. Bay S. George (Howley—Macoun);
Badger Brook (Revd. I. H. Bull—Fowler), July.

334. *E. erioceph'dus*, L. Vahl. *Lab:* Cape Chidley (Bell,
Cat. III., 347).

335. *E. Philadelphicus*, L. *Canadian or Common Fleabane.* (Rocks).

336. *E. uniflorus*, L. *One-flowered Fleabane. Lab:*
(Kohneister, Cat. II., 31); Nachvak and Cape Chidley (Bell,
Cat. III., 549); Hopedale (Weiz-Packard).

337. *E. annuus*, Pers. One solitary specimen in oats in
recently cleared land at Deer Lake (A. C. W.—Fowler). August.

338. *Erechtites hieracifolia*, Raf. *Fireweed.* Moist
places in recently burnt clearings. Very common throughout
Newfoundland and Canada, and extending west to the Saskatchewan (Cat. II., 262).

339. *Eupatorium purpureum*, L. *Joe Pye's weed.* The
Gould's and S. John's (Miss Southcott); Bay S. George (Howley
—Macoun); Topsail (Bell, Cat. III., 541). In Bay of Islands,
collected by myself; Apsey Beach (Fowler); Shoal Point and
Coal River (Trelease); New Harbour, Trinity Bay (A. C. W.),
and Cormack, in the same Bay; Salmonier River, and "a form
passing to var. *amœnum*, Gray, was collected on the Manuel's
River (R. & S.)." Wet places and sea cliffs. August and September. Prof. Macoun says (Cat. III., 541):—"Our specimens of
this species nearly all belong to the variety *maculatum*;" and he
evidently includes Dr. Bell's Topsail plant; but all my specimens
have been referred to the species itself.

340. *Gnaphalium Norvegicum*, Sumner. *Highland Cudweed. Lab:* (Torr and Gray, Cat. II., 238); Hopedale (Weiz-
Packard).

341. *G. supinum*, Vill. *Mountain or Dwarf Cudweed or
Everlasting.* Englee, White Bay (A. C. W.—Fowler). Banks.
September. *Lab:* (Morrison, Cat. II., 238); Hopedale (Weiz-
Packard).

342. *G. uliginosum*, L. *Mud or Low Cudweed.* New Harbour (A. C. W.); Summerside, B. of L., and Sandy Point, Bay S. George (A. C. W.—Fowler). Marshy meadows. Quiddi Vidi Lake, slender uliginous form, and in the bushy branched form; very abundant in burned regions, Holyrood (R. & S.). August.

343. *Hieracium pilosella*, L. *Mouse-ear Hawkweed.* S. John's (Prof. Holloway—Fowler).

344. *H. Canadense* Mx. *Canadian Hawkweed.* Harbour Deep, White Bay. In Trinity Bay, New Harbour and Bay Bull's Arm (Macoun). In and near the Bay of Islands and Little Harbour and Grand Lake (A. C. W.—Fowler); Manuel's River (R. & S.). Lakeside, roadside, and rocky river banks. August-October. *Lab.*; L'anse au Clair (A. C. W.—Fowler). August. *Flora Miq.*, damp, peaty places. Rare.

345. *H. vulgatum* Frier. " Less frequent than the preceding, and occurring in crevices of rocks by swift streams and waterfalls, Holyrood and the cataract of the Rocky River. To all appearance, indigenous. The leaves are nearly all mottled " (R. & S.). A doubtful specimen from New Harbour (A. C. W.). August. *Lab.*; (Kohmeister, Cat. II., 275); Hopedale Islands (Weiz-Packard).

346. *H. scabrum*, Mx. Bay S. George (Howley-Macoun). In Bay of Islands, Coal River, and possibly Grand Lake (A. C. W.—Fowler). September.

347. *Knautia arvensis*, Coult. *Field Knautia or Scabious* S. John's (Miss Southcott).

348. *Lactuca Canadensis*, L. *Field Lettuce.* Trinity Bay (Cormack); Harbour Breton, Fortune Bay (A. C. W.) Doubtful.

349. *L. leucophæa*, Gray. (Cat. II., 281). Bay de l'eau, Fortune Bay (Macoun), and McIver's Cove, Bay of Islands (A. C. W.—Fowler). Borders of fields. August.

350. *Leontodon autumnale*, L. *Fall Dandelion.* (HORSE DANDELION, AUGUST FLOWER). " Naturalized by becoming abundant in Newfoundland, Nova Scotia, New Brunswick and Quebec " (Cat. II., 277). Common, at any rate about Fortune

and Trinity Bay, S. John's, and the Bay of Islands (A. C. W.)
Fields and roadsides. July—September. *Flora Miq.*

351. *Matricaria inodora*, L. *Wild Chamomile.* Harbour Grace (McGill Coll. Herb., Cat. II., 253). Only on rubbish heaps, S. John's (R. & S.); Trinity Bay (A. C. W.). August.

352. *Inula Helenium*, L. *Common Elecampane.* Jackson's Arm, White Bay (Bullman). Rocky ground. August.

353. *Petasites palmata*, Gray. *Butterbur, Sweet Coltsfoot.* " Swamps and shady banks of streams from Newfoundland and Labrador to Rocky Mountains (Richardson, Cat. II., 260); White Bay (Revd. S. L. Andrewes—Macoun). *Lab:* Hopedale (Weiz-Packard).

354. *Prenanthes alba*, L. *White Lettuce.* *(PIGROOT). (Cormack, Cat. II., 282). (Reeks). *Flora Miq.*, very common in bushes. August.

355. *P. serpentaria*, Pursh. *Rattlesnake Root.* Trinity Bay (Cormack), and New Harbour Brook (6) (A. C. W.); S. John's (Miss Southcott and Lady Blake), and Bay D'Espoir in Hermitage Bay: all named by Prof. Macoun. Var. *Nana*, Gray. Holyrood and S. John's (R. & S.). Rocky hillsides, 6 inches to $2\frac{1}{2}$ feet high. August. *Lab:* L'anse au Clair (A. C. W.—Fowler). September.

356. *P. altissima*, L. *Tall White Lettuce.* (Reeks); (Cormack, Cat. II., 282).

357. *Onopordon Acanthium*, L. *Cotton or Scotch Thistle.* Harbour Grace (Clift).

358. *Rudbeckia hirta*, L. *Coneflower.* S. John's and Holyrood, not yet abundant, (R. & S.). Near meadows, Bay of Islands (A. C. W.—Fowler); Tiddleton, near Conception Bay (Clift—Macoun). Fields. August.

359. *Senecio aureus*, L. *Golden Groundsel, Common Ragwort.* " From Newfoundland and Labrador to the Rocky

---

* *The Flora Miqueloncnsis* remarks on this plant,—"Pigs are very fond of the root known as the plant under the name of 'Mountain Turnip.' It gives to the flesh an excellent flavour."

Mountains and the Pacific." (Cat. II., 264); (Reeks); S. John's (Professor Holloway—Fletcher); Salmonier and near Placentia Junction (R. & S.); near Badger Brook (Bull—Fowler); by myself in the Bay of Islands at Benoit's Cove, McIver's Cove and other places (Fowler & Coville); White Bay (Macoun). Bogs, June and July. *Lab*: found by myself at Capstan Island, L'anse au Clair, and Forteau, in the Straits of Belle Isle (Fowler and Eaton). July—August.

Var. *obovatus*, T. & G. Bay East River, Hermitage Bay (Howley-Macoun).

Var. *borealis*, T. & G. *Lab*: (Gray, Cat. II., 265); Nachvak (Bell, Cat. III., 354); Hopedale Islands (Weiz-Packard).

Var. *discoideus*, Hook. *Lab*: (Pursh, Cat. III., 265); by myself at Forteau (Eaton). July.

Var. *Balsamitæ*, T. & G. Holyrood and Exploit's River (R. & S.); Chimney Cove, Birchy Cove, and on the Bay of Islands (Fowler and Robinson). June. *Lab*: at Forteau and L'anse au Mort (Fowler), and Long Point, Hamilton Inlet (Macoun); (Butler). Swamps. June—August.

360. *S. frigidus*, Less. "Newfoundland (?) and Labrador" (Gray, Cat. II., 267).

361. *S. palustris*, Hook. *Marsh Groundsel or Fleawort*. *Lab*: Indian Harbour, North (Revd. W. How—Macoun).

Var. *congestus*, Hook. Battle Harbour (Bull); a northern form at Seal Islands (A. C. W.); both determined by Prof. Macoun. Marshes. August.

362. *S. Pseudo-Arnica*, Less. *False Arnica*. Newfoundland and Labrador (Hooker, Cat. II., 267); Harbour Breton (Macoun), and Sandy Point, Bay S. George (Fowler); several places about the entrance to Bay of Islands (A. C. W.). *Lab*: (Butler); Hopedale Islands (Weiz-Packard); Battle Harbour (Bull—Macoun); by myself at L'anse au Mort (Fowler); and further north about Sandwich Bay. Sea beach and sandy places. July and August. *Flora Miq.*, dry and stony places. July, August.

363. *S. vulgaris*, L. *Common Groundsel*. "Newfoundland and Labrador and Hudson Bay (Hooker, Cat. II., 263); appears

to be common everywhere in and about gardens;" sandy shore, strange to say, Flat Bay (Bell).

364. *S. sylvaticus*, L.   Railway ballart, Whitbourne; abundant (R. & S.).   August.

365. *S. Jacobæa*, L.   *Common Ragwort.*   S. John's (R. & S.)   Roadsides,   August.

366. *Sonchus asper*, Vill.   *Spiney Sow Thistle.*   New Harbour (A. C. W.), and Bay d' Espoir, Hermitage Bay (Mrs. Gallop-Macoun).   August.

367. *S. arvensis*, L.   *Corn or Field Sow Thistle.*   "Abundant along roadsides and in fields from Newfoundland throughout the Atlantic Provinces and Quebec " (Cat. II., 283); New Harbour (A. C. W.).   Gravel banks in Salmonier River, exclusively with native plants, as if indigenous (R. & S.)   August.

368. *S. oleraceus*, L.   *Common Sow Thistle.*   Fields, Placentia (R. & S.).   In Trinity Bay, at Heart's Content (Miss Southcott); and at Rawdon and New Harbour (A. C. W.). " Naturalized from Newfoundland to Manitoba and B. Columbia " (Cat. II., 283).   August.

369. *Solidago rugosa*, Mill.   *Tall Golden Rod.*   Holyrood. A smoothish form was collected in open woods near S. John's R. & S.) ; at Harbour Breton (Macoun) ; in Bay of Islands, at Little Harbour, and Lark Harbour (Fowler) ; and at Benton, Bonavista Bay (Trelease) by myself.   Fields and wet woods. August, September.

370. *S. arguta*, Ait.   (Reeks).

371. *S. bicolor*, L.   v. *concolor*, T. & G.   White Bay (Bullman) ; in Bay of Islands, at Apsey Beach, and Shoal Point (Fowler).   Two other specimens are referred to this plant by Dr. Robinson, one from Goose Arm (called *S. nemoralis*, " low canescent form " by Dr. Trelease), and the other from Coal River (said to be *S. humilis*, by Prof. Fowler).   Sea cliffs and woods. August.

372. *S. Canadensis*, L.   *Canadian or Common Golden Rod.* (Reeks).   Trinity Bay (Cormack) ; S. John's (Miss Southcott) ;

Bay S. George (Mr. Howley-Macoun); Harbor Breton, Fortune
Bay (A. C. W.), also by myself at Goose Arm; Bay of Islands
(Fowler). *Lab*: L'anse au Chair (A. C. W.—Fowler). Woods
August, September.

373. *S. cæsia*, L. ?
Var. *flexicaulis*, Hook. } I don't see the authority for these.

374. *S. humilis*, Pursh. (Herb. Banks, Cat. II., 213); col-
lected by myself at Long Cove, Trinity Bay, and at Harbour
Breton; in Fortune Bay (Macoun); at Gander and Exploit's
Rivers, Notre Dame Bay (Drummond, E. E.—Macoun). *Lab*:
(McGill Coll. Herb., Cat. II., 213—Ungava, just outside our
Northern limit).

375. *S. juncea*, Ait. ? S. John's (Miss Southcott) [?].

376. *S. latifolia*, L. Trinity Bay (Cormack); at Lark
Harbour (Fowler). Woods, September. .

377. *S. multiradiata*, Ait. Harbour Breton (A. C. W.).
Lab: collected by myself at Mullin's Cove, Hamilton Inlet, and
Independent Sandwich Bay (Macoun), and Forteau, in the Straits
(Fowler). July, August.

378. *S. serotina*, Ait. var. *gigantea*, Gray. *Gigantic
Golden Rod*. "Borders of thicket and low grounds. Common
throughout Canada, Newfoundland, Nova Scotia, and westward
to the Pacific".(Cat. II., 216); New Harbour (A. C. W.).

379. *S. squarrosa*, Muhl. *Ragged Golden Rod*. *Flora Miq.*,
Chappean Hill. Common.

380. *S. Terræ-Novæ*, T. & G. (Pylaie, Cat. II., 215), Whitbourne. "Clearly a more corymbosely branched form of *S.
uliginosa*, towards which intergradations were found near the
Exploits River" (R. & S.). Bogs. August. *Flora Miq.*, Chappean Hill. Common.

381. *S. puberula*, Nutt. White Bay (Bullman). August.

382. *S. nemoralis*, Ait. White Bay (Bullman). Banks.
August. (Vide *S. bicolor*, var. *concolor*, No. 371, above).

384. *S. macrophylla*, Pursh. S. John's (Miss Southcott and
R. & S.); by myself at New Harbour, Trinity Bay, Harbour

Breton in Fortune Bay (Macoun); Lark Harbour (Coville), and Goose Arm (Robinson); Bay of Islands and at Benton, Bonavista Bay (Covil'e). *Lab:* (McGill Coll. Herb., Cat. II., 212); Ford's Harbour (Bell, Cat. III., 543); by myself at Forteau and L'anse au Clair (Fowler) in the Straits, and north of this at Battle Harbour, Deep Water Creek, and Venison Tickle (Macoun). Woods. August—October.

385. *S. uliginosa*, Nutt. *Swamp Golden Rod.* "Newfoundland, to, and beyond the Rocky Mountains in the wooded country" (Cat. II., 214); S. John's (Miss Southcott); by myself at New Harbour and Bay Bull's Arm; Trinity Bay (A. C. W.); in Bay of Islands at Shoal Point (Fowler); and near Riverhead (Robinson); and at Benton, Bonavista Bay (F.); Exploit's River, etc., (R. & S.). Wet places and woods. August.

386. *S. Virgaurea*, L. var. *alpina*, Bigel. Nipper's Harbour and Belt Cove, Notre Dame Bay (Bull—Macoun). *Lab:* Ford's Harbour and Nachvak (Bell, Cat. III., 543); Hopedale (Weiz-Packard).

387. *S. sempervirens*, L. Harbour Breton (A. C. W.).

388. *Tanacetum vulgare*, L. *Tansy.* (Reeks). Trinity Bay and Bay of Islands, here and there. *Flora Miq.*, near dwellings. August, September.

389. *Taraxacum officinale*, Weber. *Dandelion* (DUMBLE-DOR). Appears to be common throughout Newfoundland about settled places, and at Battle Harbour, and some other places on the Labrador. *Flora Miq.*, common. July.

Var. alpina, Koch. *Lab:* not uncommon along the Labrador coast. Flowers usually very large (W. A. Stearns). Labrador to B. Columbia (Gray, Cat. II., 289). Rocky soil at Nachvak and Nain (Bell, III., 558); Hopedale (Weiz); and Caribou Islands (Butler-Packard).

390. *Tussilago Farfara*, L. *Coltsfoot.* S. John's (Holloway —Macoun). Lower Brook, Bay of Islands (introduced from Nova Scotia).

The *Flora Miquelonensis* notes,—" Erigeron Canadensis, L.,
Solidago Canadensis, L., Aster tripolium, L., Artemisia borealis,
L., Carduus nutans, L., Cineraria carnosa, de la Pel., Hypochæris
radicata, mentioned by Gunthier, have not been found" (by us).

LIII.—LOBELIACEÆ. *Lobelia Family.*

391. *Lobelia Dortmanna,* L.   *Water Lobelia.* West of
Random, Trinity Bay (Cormack), and in the same Bay, Green's
Harbour and New Harbour (A. C. W.); Placentia (Lady Blake);
Quidi Vidi Lake (R. & S.). Shallows, ponds and brooks. *Flora
Miq.,* very common.

392. *L. Kalmii,* L.   *Kalm's Lobelia.* In Bay of Islands,
at Goose Arm and Middle Arm, in Bay S. George, at Seal Rock,
Sandy Point (A. C. W.,—Fowler). Wet quarry places and bogs
August.

LIV.—CAMPANULACEÆ. *Campanula Family.*

393. *Campanula uniflora,* L. *Lab:* Arctic regions from
Labrador to Aleutian Islands (Gray, Cat. II., 287); Nachvak and
Cape Chidley (Bell, Cat. III., 559); Hopedale (Weiz-Packard).

394. *Campanula rotundifolia,* L. *Rock Bellflower, Hare-
bell.* Petty Harbour (Bell, Cat. II., 559); reported from S.
John's, Bay S. George, found by myself in Trinity and Fortune
Bays, White Bay and Bay of Islands. *Lab:* common (Butler
and Stearns, Cat. II., 288); Battle Harbour and several places in
the Straits (A. C. W.). *Flora Miq.,* abounds in the fields and in
the damp portions of the island. Cliffs and rocky and sandy
places. July, August.

Var. *arctica,* Lange. This is the one few-flowered forms
and ranges from Canada and Labrador to the arctic regions"
(Gray, Cat. III., 560); Nachvak and Cape Chidley (Bell, Cat.
III., 560); Hopedale (Weiz), and L'anse Amour and Caribou
Islands (Butler and Martin—Packard).

C. Scheuchzeri, Vill. Newfoundland, Labrador and Alaska
(Gray, Cat. II., 287); New Harbour (A. C. W.).

Var. *heterodoxa*, Gray.   Near the coast on western side of Newfoundland (Pylaie, Cat. II., 288).

These Professor Macoun (Cat. III., 560) refers to *C. rotundifolia*, L.

LV.—VACCINIACEÆ.   *Huckleberry Family.*

395.   *Chiogenes hispidula*, T. & G.   *Creeping Snowberry.* (MAIDENHAIR, CAPILLAIRE) seems to be common and widespread in most woody parts of the country and on the Labrador (A. C. W.), so Drummond in Cat. II., 351.   "Damp mossy woods, creeping over logs." *Flora Miq.*, very common.   May—July.

396.   *Gaylussacia dumosa*, T. & G.   *Dwarf or Pale Huckleberry* (Gray, Cat. II., 289) ; Whitbourne (R. & S.) ; Little Bay, Fortune Bay.   Edge of woods.   August.

397.   *G. resinosa*, T. & G.   *Black Huckleberry.*   (BLACK HURTS).   (Reeks) : (Cat. II., 289 : rocky or sandy woodland, or swamps) ; by myself at New Harbour (Trinity Bay), Long Harbour (Fortune Bay), and at Little Harbour near Bay of Islands (Macoun and Fowler).   Wet places.   July.

398.   *Oxycoccus vulgaris*, Pursh.   *Common or Small Cranberry.*   (MARSHBERRY).   (Reeks).   Very common in bogs, it would appear, throughout Newfoundland and Labrador (A. C. W.).   *Lab :* Hopedale (Weiz), and Caribou Island (Butler-Packard).   June—August.

399.   *O. macrocarpus*, Pursh.   *Large American Cranberry,* (CRANBERRY, BEARBERRY and BANKBERRY).   Bogs, and especially on the margins of ponds and small lakelets in the soft mud. Newfoundland, Anticosti, Nova Scotia, etc., to Thunder Bay Macoun, Cat. II., 293) ; West of Random (Cormack), and New Harbour (A. C. W.) in the same neighbourhood ; Cod Roy River (Bell), and Bay of Islands.   Much less frequent than the last ; said to be common about Lamaline and Lawn in Burin district, there called Bankberry.   *Lab :* by lakelets along the coast (Abbé (Brunot : Packard).   *Flora Miq.*, says of this and the last " barrens, hills, dry or damp places, almost everywhere, very common."   June—August.

400. *Vaccinium Pennsylvanicum, Lam. Common Low or Early Fruiting Blueberry or Whortleberry. (LOW BUSH HURTS). Very abundant on burnt tracts (R. & S.); seems to be about our most common whortleberry. I have it from White Bay, Notre Dame Bay, Bay S. George, Trinity Bay and Bay of Islands. Flora Miq., very common; open woods and barrens. June to August.

Var. angustifolium, Gray. GROUND HURT. On the Labrador, TOBACCO HURT. Rocky hills, Placentia, infrequent; Salmonier (R. & S.); Trinity Bay and Bay of Islands (A. C. W.). Lab: (Gray, Cat. II., 290); hillsides and Caribou Island (Butler); Nain (Lundbery--Packard); Snack Cove, Sandwich Bay (A.C.W.) Flora Miq.: very common. June—August.

401. V. cœspitosum, Mx. Dwarf or Tufty Bilberry or Blueberry. (SUGAR HURT—Labrador), (Reeks); Lab: Hopedale (Weiz), and Belles Amours, and on Caribou Islands (Butler—Packard); (at Snack Cove, near Sandwich Bay, and Cape Charles (A. C. W.). Hillsides. July.

402. V. Canadense, Kalm. Canadian Blueberry. (Reeks); Bay S. George, White Bay and Bonne Bay (Bullman); Harbour Breton, Fortune Bay (A. C. W.). July.

403. V. corymbosum, L. Swamp Blueberry. (Reeks); swamps and low woods from Newfoundland to Western Ontario (Gray; Cat. II., 290); S. Paul's Bay, N. W. coast (Bullman). Wet places. July.

404. V. Vitis-Idœa, L. Cowberry, Red Whortleberry (PARTRIDGE BERRY, †REDBERRY). Very abundant from the Atlantic to the Pacific, except Southern Ontario and the prairie regions (Hook, Cat. II., 292). Appears to be abundant and widely distributed throughout Newfoundland and the Labrador.

---

* The Vaccinium family (excepting V. Vitis Idœa, L.) is generally called by our people on the east coast, "hurts;" on the west coast, "blue berries."

†A dear old friend of mine, writing from the neighbourhood of Sandwich Bay, Labrador, told me, a few years ago, that she and the three girls had that fall gathered and sold 40 gallons of "bakeapples" (Rubus Chamæmorus), and 28 gallons of "redberries."

Flora Miq., abounds in the peaty plains and also in the dry parts of the island. June—July.

405. *V. uliginosum*, L. *Mountain or Bog Blueberry* (GROUND HURT), (Cat. II., 291). In mountain bogs and exposed shores below. From Newfoundland, Labrador, etc., thence westward to the Pacific, and northward to the Arctic Sea. Flat Bay (Bell); several places in Trinity Bay and Bay of Islands (A. C. W.); S. John's and Holyrood (R. & S.). *Lab:* at Blanc Sablon (Strait), Deep Water Creek, Seal Islands, and Hamilton Inlet (A. C. W.). Common on the coast at Nain, Ford's Harbour and Nachvak (Bell-Packard). *Flora Miq.*, very common. June —July.

406. *V. ovalifolium*, Smith. (BLUEBERRY HURT, MAZZARD). Frenchman's Cove, B. of I., in woods (A. C. W.—Robinson). *Lab:* *West S. Modest, in the Strait of Belle Isle. June.

LVI.—ERICACEÆ.   *Heath Family.*

407. *Arctostaphylos Uva-Ursi*, Spreng. *Red Bearberry* (INDIAN HURT and HARDBERRY, in Hermitage Bay). *(Richardson)* rocky or sandy soil from Newfoundland to the Pacific, and north to Fort Franklin, Lat. 64° (Richardson, Cat. II., 295); Trinity Bay (Cormack and A. C. W.); Harbour Breton (Fortune Bay) and Bay de Verde, East coast, and Chimney Cove, Bay of Islands (A. C. W.) Also, Sampson's Island, Notre Dame Bay, Flat Bay Brook (Bell). Sea cliffs and rocky banks. June.

408. *A. alpina*, Spreng. *Alpine or Black Bearberry.* Trinity Bay (Cormack); by myself at Branet and Harbour Breton (Fortune Bay), Bay de Verde, and Swan Island, Notre Dame Bay, Great Cod Roy River (Bell). *Lab:* (McGill Coll. Herb., Cat. II., 294); Ford's Harbour and Cape Chidley (Bell: Cat III., 561); Hopedale (Weiz-Packard): by myself at Battle Harbour and L'anse au Loup. *Flora Miq.*, common. Hill tops. June.

---

*An intelligent resident of this place informed me that the fruit made good wine ; which would imply that it was fairly plentiful, but I have only met with it once, and then in very small quantity.

409. *Andromeda Polifolia*, L. *Marsh Andromeda, Wild Rosemary.* (Reeks) ; common about Fortune and Trinity Bays, Chimney Cove and Little Harbour, Bay of Islands (A. C. W.) ; S. John's (Miss Southcott) ; Exploits River, and near Whitbourne (R. & S.) ; near Flat Bay, Bay S. George (Bell) ; *Lab :* Hopedale (Weiz), Caribou Island (Butler—Packard), Indian Harbour, near Battle Harbour, Square Islands, and other places (A. C. W.). *Flora Miq.*, very common. Bogs. June—August.

410. *Bryanthus taxifolius,* Gray. Mountains at Great Cod Roy River (Bell). *Lab :* (Morrison, Cat. II., 299) ; Nain, Nachvak, and Ford's Harbour (Bell, Cat. III., 562) ; Hopedale (Weiz-Packard) ; Battle Harbour and Seal Islands, and L'anse au Clair (A. C. W.). June—August.

411. *Chimaphila umbellata*, Nutt. *Prince's Pine* (Reeks).

412. *Cassandre calyculata*, Don. *Leatherleaf.* (Cat. II., 296). Common in Trinity and Fortune Bays and Bay of Islands (A. C. W.) ; near Flat Bay (Bell) ; Exploits River and Salmonier (R. & S.). *Lab :* (Cat. II., 296) ; Battle Harbour (A. C. W.) ; borders of lakelets and swamps along the coast (Hooker) ; Square Islands (B. P. Mann—Packard). *Flora Miq.*, very common. April—July.

413. *Calluna vulgaris,* Salisb. *Heath or Heather.* Near Caplin Bay, Ferryland, Renews ; S. Mary's Bay and Trepassey Bay (Cormack and Lawson, Cat. II., 298) (Reeks).

414. *Cassiope hypnoides,* Don. *Moss-like Cassiope, Moss-plant.* *Lab :* (Morrison, Cat. II , 296) ; Nain and Cape Chidley (Bell, Cat. III., 562) ; Hopedale (Weiz-Packard).

415. *C. tetragona,* Don. A specimen in the museum in S. John's, named *Menziesia Polifolia,* Professor Macoun thinks may be this. *Lab :* (Kohneister and Douglas, Cat. II., 297). Abundant along coast at Nain (Bell, Cat. III., 562) ; Hopedale (Weiz-Packard).

416. *Epigæa repens*, L. *Trailing Arbutus, May Flower.* (Reeks). Near Flat Bay Brook (Bell) ; Bonne Bay, common (Bullman) ; also at the Bay of Islands ; Little River (Burgess) ;

Hermitage Bay : abundant at Rose Blanche (Revd. G. A. Field).
Woods and thickets.   April—June.

417. *Gaultheria procumbens*, L. *Boxberry, Tea-berry, or
Partridge Berry* (MOUNTAINEER TEA). (Cat. II., 295) ; near
Harbour Breton (A. C. W.) ; S. Paul's Bay, West coast (Bullman);
(Reeks). *Flora Miq.*, common.   June—September.

418. *Kalmia glauca*, Ait. *Pale or Swamp Laurel* (GOLD
or GOULDWITHY) (Reeks). Common in peat bogs throughout
the country apparently (A. C. W.). *Lab :* hillsides and swamps,
Caribou (Butler-Packard) ; Battle Harbour, and a few other
places ; not so common where I have been on the Labrador as
the next (A. C. W.). *Flora Miq.*, abundant ; " one of the first to
flower ; it is also found in flower in some places even in August
and September,

419. **K. angustifolia*, L. *Sheep-laurel or Lamb-hill* (Cat.
II , 300). Apparently as frequent and widespread as the last
(A. C. W.). *Lab :* (Cat. II., 300). West S. Modest, Chatham,
and Battle Harbour (A. C. W.). *Flora Miq.*, very common.
Wet and rocky places.   July—September.

420. *K. latifolium*, L. *Mountain Laurel, Calico Bush.*
Lab : reported as being found in ravines and near ponds in the
interior up Salmon River, and on Esquimaux Island (Stearns,
Cat. II., 300).

421. *Loiseleuria procumbens*, Desv. *Alpine or Trailing
Azalea* (MAY FLOWER and WHITE FLOWER in Hermitage Bay).
(Morrison, Cat. II., 298) ; Flat Bay Brook (Bell) ; Pustelbrough,
Hermitage Bay (Revd. H. G. Bishop) ; Chimney Cove, B. of I.
(A. C. W.). *Lab :* (Morrison) ; hillsides, Caribou (Butler, Cat.
II , 298) ; Battle Harbour and Seal Islands (A. C. W.) ; Hope-
dale (Weiz) and Ford's Harbour (Bell — Packard). *Flora Miq.*,
dry places ; not common.   Hills.   June—September.

422. *Ledum palustre*, L. (CRYSTAL TEA, Labrador) (Reeks).
Newfoundland and Labrador, and through the Arctic region to

'Called everywhere, like K. glauca, " Gouldwithy." This appears in Trinity Bay
to be taken as the earlier-flowering K. glauca, in its second bloom.

Alaska and Aleutian Islands (Gray, Cat. II., 301); (Cormack). *Lab:* Seal Islands, Pack's Harbour and Snack Cove (A. C. W.); Hopedale (Weiz), and Nachvak, Ford's Harbour and Cape Chidley (Bell—Packard) (Cat. III., 563). *Flora Miq.* says of this and the next—very common. Bogs. July.

423. *L. latifolium*, Ait. *Labrador Tea* (INDIAN or LABRADOR TEA). Peat bogs and marshes from Labrador, Newfoundland and westward to the Pacific (Cat. II., 301). Appears to be frequent in suitable habitats. *Lab:* common on hills Caribou Islands (Butler); Hopedale (Weiz-Packard). July—August.

424. *Monesis uniflora*, Gray. *One-flowered Pyrola* or *Wintergreen* (SWEET FLOWER, in White Bay). Very common in shady or mossy woods, from Labrador, Newfoundland, etc., westward to the Pacific and northward to Lat. 64° (Hooker, Cat. II., 306); Whitbourne, rare (R. & S.); fairly common in Trinity and Fortune Bays and Bay of Islands (A. C. W.—Macoun and Fowler); Great Cod Roy River (Bell); S. John's (Miss Southcott) and Ferryland. *Lab:* in damp and shady places (Butler); Turner's Head (Hamilton Inlet); Venison Tickle, S. Michael's, and further south at L'anse au Clair (A. C. W.); Hopedale (Weiz-Packard). *Flora Miq.*, found in groups of 12 or 15 individuals in damp peaty places, but always rare. July—August.

425. *Pyrola secunda*, L. *One-sided or Serrate Pyrola*, or *Wintergreen.* Rich woods throughout Canada, from Newfoundland, etc., to the Pacific, and far northward on the Mackenzie (Hooker, Cat. II., 304); S. John's (Lady Blake and R. & S.); (Reeks); White Bay, Harbour Breton and New Harbour (A.C.W.). *Lab:* in the Straits of Belle Isle, at Forteau, L'anse au Mort, and Capstan Island (Eaton and Fowler). *Flora Miq.*, common. August—September.

Var. minor, Gray. Chimney Cove, Bay of Islands (A.C.W.). *Lab:* peaty bogs and mossy swamps, from Labrador to Alaska (Gray, Cat. II., 304); Hopedale (Weiz-Packard); Forteau and L'anse au Mort (A. C. W.—Fowler). Woods. August.

426. *P. minor*, L. *Smaller Pyrola.* Near Cairn Mountain, Flat Bay (Bell); (Reeks); New Harbour, in Trinity Bay (A.C W.).

Lab: cold woods (Morrison, Cat. II., 303): Battle Harbour and neighbourhood, and in the Strait, Forteau and Blanc Sablon (A. C. W.—Fowler); Hopedale (Weiz-Packard).   July, August.

427. *P. chlorantha*, Swartz. *Green-flowered or Small Pyrola*. Rather dry or sandy woods, generally under conifers, from Newfoundland, Labrador, etc., westward to the Rocky Mountains, and northward to Bear Lake (Richardson and Gray, Cat. II., 304); (Reeks); Lark Harbour and Coal River, Bay of Islands (A. C. W.—Fowler); Exploits and Gander River (Drummond, C. E.—Macoun); S. John's (R & S.). *Lab*: (Morrison—Packard). July—August.

428. *P. rotundifolia*, L. *Round-leaved Wintergreen*. Sandy or dry woods, in swamps, and on mountain tops, from the Atlantic to the Pacific, and northward to the Arctic regions (Hooker, Cat. II., 305); Harbour Grace (Miss Trapnell—A. H. MacKay); near Cairn Mountain, Flat Bay (Bell); S.John's (Lady Blake); Sphagnum Swamp, Manuel's (R. & S.). *Lab*: in swamps, Amour (Butler). August.

Var. pumila, Hook. *Lab*: from Labrador to the Mackenzie River (Gray, Cat. II., 305). Quite common along the Labrador coast (Butler, Cat. II., 503); Battle Harbour (A. C. W.); Hopedale (Weiz-Packard). September.

Var. incarnata, D. C. *Lab*: Battle Harbour (A. C. W.). August. A doubtful specimen.

429. *Rhododendron Rhodora*, Don. *Pink Rhodora. False Honeysuckle* (BULL'S EYE, BULL'S TONGUE). Cool bogs and open peaty places, from Newfoundland, Labrador, etc., westward to the vicinity of Montreal (Maclagan, Cat. II., 302); Flat Bay, and on Cairn Mountain, white specimen (Bell); S. John's and Exploit's River, abundant (R. & S.); appears to be common and widely diffused (A. C. W.). *Lab*: hillsides, Caribou Islands (Butler-Packard). *Flora Miq.*, common. June—July.

430. *R. Lapponicum*, Wahl. *Lab*: (Morrison, Cat. II., 302); Nachvak (Bell); Hopedale (Weiz); on a hill top Belles Amours (Butler—Packard).

LVII.—MONOTROPACEÆ. *Pipewort Family.*

431. *M notropa uniflora,* L. *Indian Pipe. Corpseplant* (GHOSTPLANT OR GHOSTFLOWER). New Harbour and B. of I., Fortune Bay (A. C. W.); near Cairn Mountain, Flat Bay (Bell); (Reeks); S. John's (Miss Southcott and R. & S. Messrs. Robinson and Schrenk remark,—" in woods near the Exploits River a small form was found, which, although agreeing as to anther and stigma with *M. uniflora,* had flowers in size just intermediate between this and *M. Hypopitys.* In drying, also, these plants have assumed an intermediate color between the black of the former species and the tawny color of the latter." Bear's Harbour, Parson's Pond, Bonne Bay (Bullman). Woods. August.

452. *Hypopitys lanuginosa,* Nutt. *Yellow or Pine Birdsnest or Pinesap* (Reeks); in woods near Exploits River (R. & S.); Great Cod Roy River (Bell); White Bay (Bullman). Woods. July, August.

LVIII.—DIAPENSIACEÆ. *Diapensia Family.*

433. *Diapensia Lapponica,* L. *Northern Diapensia* (MOSS LILY, GROUND IVORY FLOWER). Western Head, Harbour Breton and Conne in Fortune Bay, and near Rantem, Trinity Bay (A. C. W.). *Lab:* (Morrison), common on hill tops at Caribou (Butler, Cat. II., 308); Nain, Ford's Harbour and Cape Chidley (Bell, Cat. III., 564); Hopedale (Weiz—Packard). *Flora Miq.,* very common. Hills. June, July.

LIX.—PLUMBAGINACEÆ. *Leadwort Family.*

434. *Armeria vulgaris,* Willd. *Common Thrift, Sea Pink.* (Reeks); Coal River (A. C. W.—Fowler). *Lab:* Labrador, Newfoundland, and N. W. America, and in the barren country of the interior (Hooker, Cat. II., 309); Nain, Nachvak, and Cape Chidley (Bell); Hopedale (Weiz) (Cat. III., 564, and Packard). Hills and sandy plains. July.

435. *Statice Limonium,* L. Var. *Carolinianum,* Gray. *Sea Lavender, Marsh-Rosemary.* (Miss Brenton, Cat. II., 308); (Reeks); *Lab:* (Gray, Cat. II., 308).

LX.—PRIMULACEÆ.   *Primrose Family.*

436. *Anagallis arvensis,* L.   *Common or Scarlet Pimpernel.*
Harbour Grace (McGill Coll. Herb., Cat. II., 315).   *Flora Miq.,*
introduced into cultivated places.

437. *A. tenella,* L.   *Flora Miq.,* fields; common.

438. *Glaux maritima,* L.   *Sea Milkwort, Black Saltweed.*
Salt marshes along the coast of the Atlantic, from Newfoundland
and Labrador to the Coast of Maine (Cat. II., 315).

439. *Lysimachia stricta,* Ait.   *Wood Loosestrife.*   Newfoundland to the Saskatchewan (Gray, Cat. II., 314) ; moist
ground, Whitbourne (R & S.) ; Trinity Bay (Cormack), and by
myself in the same Bay at Spreadeagle ; Ferryland (Revd. R.
Temple—Fowler) ; near Brigus (Miss Trapnell—MacKay). Wet
places. August. *Flora Miq.*

440. *L. Nummularia,* L.   *Moneywort.*   Harbour Grace
(McGill Coll. Herb., Cat. II., 314).

441. *Primula Mistassinica,* Mx. (Reeks ; common) ; Middle Arm, Grand Lake, and several other places in and about Bay
of Islands (A. C. W.—Fowler) ; Flat Bay (Bell) ; Batteau barrens,
N. W. coast (Bullman). *Lab :* Bonne Esperance and neighbour-
ing islands at Forteau (Butler, Cat. II., 309) ; Hopedale (Weiz-
Packard) ; Battle Harbour (A. C. W.).   Rocky and most exposed
places.   June—August.

442. *P. farinosa,* L. *Bird's Eye Primrose* (SALMON FLOWER).
(Reeks ; common) ; crevices of rocks, Port à Port (Bell) ; Englee,
White Bay, Middle Arm, and Chimney Cove, and Coal River
(A. C. W.—Fowler) ; Flower's Cove, in the Straits (Spence).
Lab : crevices of rocks and exposed points along the sea, lakes
or rivers (Butler), Hopedale (Weiz) ; Caribou Island and L'anse
Amour (Butler - Packard) ; Sandwich Bay, Battle Harbour,
Long Point (Hamilton Inlet), and L'anse au Clair (A. C. W.—
Macoun and Fowler).   July.

443. *P. Egaliksensis,* Hornem. *Lab :* Northern (Cat. III.,
564).

444. *Samolus Valerandi*, L. Var. *Americanus*, Gray. *Water Pimpernel* (Reeks).

445. *Trientalis Americana*, Pursh. *Chickweed Winter-green*, *Star Flower*. Appears to be quite common everywhere, chiefly in damp grassy woods, in Newfoundland and on the Labrador (A. C. W., Cat. II., 313, and Packard). June—August.

LXI.—OLEACEÆ. *Olive Family.*

446. *Fraxinus Americana*, L. *White Ash* (Reeks); a very rare tree ; only in the country surrounding S. George's and Port a Port Bays (Prof. Howley in Geological Survey Report, p. 44).

447. *F. pubescens*, Lam. *Red or River Ash* (Reeks: common) ; (Howley).

448. *F. sambucifolia*, Lam. *Black or Swamp Ash.* Humber River, quite abundant, and Deer Lake, in and about Bay of Islands (Bell).

LXII.—APOCYNACEÆ. *Dog Bane Family.*

449. *Apocynum cannabinum*, L. *Indian Hemp.* Badger's Brook (Revd. I. H. Bull—Fowler). August.

450. *A. androsæmifolium*, L. Exploits River (R. & S.). Open woods. August.

LXIV.—GENTIANACEÆ. *Gentian Family.*

451. *Bartonia tenella*, Muhl. *Screwstem.* Open woods. (Gray, Cat. II., 327).

452. *Bartonia*, sp.* *(Centaurella Moseri, Steud. & Hochst.).* A plant which appears to represent, at least in part, this rare and poorly understood species, was discovered in a small bog near Holyrood (Robinson and Schrenk ; for their further remarks on this plant see their " notes "). August.

453. *Gentiana crinita*, Fræl. *Blue-fringed Gentian* (Reeks).

Bartonia iodandra, Robinson. Botanical Gazette, 1898, July, p. 47.

454. *G. quinqueflora*, Lam. Parson's Po.d, Bonne Bay (Bullman). Dry hillsides. August.

455. *G. propinqua*, Richards. *Lab:* (Gray, Cat II , 322). On hillsides at Amour and lowlands at Bonne Esperance (Stearns —Packard). Prof. Macoun notes, however, that this is more likely to be the next.

456. *G. Amarella*, L. var. *acuta*, Hook. *Autumnal or Small-flowered Gentian.* (Reeks: common in short grass); Chimney Cove, Bay of Islands (A. C. W.—Robinson). *Lab:* (Gray, Cat. II., 322); Hopedale (Weiz), and Caribou Island (Butler—Packard); by myself in the Strait, L'anse au Clair and Forteau (Fowler). August.

457. *G. serrata*, Gunner. *Shorn or Smaller-fringed Gentian.* Wet grounds, by streams and on rocks (Gray, Cat. II., 321); Englee (Revd. R. Temple—Fowler).

458. *G. Andrewsii*, Griseb. (Reeks).

459. *G. nivalis*, L. *Small Alpine Gentian.* *Lab:* collected by the Moravian Brethren (Gray), and Hopedale (Weiz--Packard).

460. *Halenia deflexa*, Griseb. *Spursed Gentian. Felwort.* (Reeks); Conche (N. E. coast), New Harbour and Bay of Islands or Chimney Cove (A. C. W.—Macoun and Fowler). *Lab:* Forteau Bay (Miss Brodie); on hillsides at L'anse Amour, and lowlands at Bonne Espérance (Stearns); Caribou Islands (Butler —Packard); Bluff Head (Hamilton Inlet), Capstan Islands and Forteau in the Strait (A. C. W.—Fowler). Hills. August.

Var. Brentoniana, Gray. Harbour Grace (Cat. II., 326). Rocky hills, S. John's (R. & S.) August.

461. *Menyanthes trifoliata*, L. *Buck or Bog Bean.* Quite common in bogs, swamps and slow-flowing streams, from Labrador, Newfoundland, etc., to the Pacific, and northward to Sitka (Cat. II., 327) (Reeks): West of Random (Cormack), and at New Harbour in the same neighbourhood (A. C. W.); also at Hermitage Bay, Fortune Bay, Bonne Bay and Bay of Islands (A.C.W.) *Lab:* Hopedale (Weiz), and Caribou Island (Butler—Packard); Holton (A. C. W.). *Flora Miq.*, common. June.

462. *Pleurogyne rotata*, Griseb. (Cat. II., 325); Englee (A. C. W.). *Lab :* on the flat at Caribou and low lands at Bonne Espérance (Stearns—Packard); Battle Harbour, and at Sandwich Bay, and Hamilton Inlet. August.

Var ? Harbour Breton (A. C. W.—Macoun).

463. *P. carinthiaca*, Griseb. Var. *pusilla*, Gray. Conche, sea beach (A. C. W.—Fowler). *Lab :* (Hooker, Cat. II., 325); Battle Harbour and neighbourhood (A. C. W.) August—September.

LXVII.—BORAGINACEÆ. *Borage Family.*

464. *Cynoglossum officinale*, L. *Common Hound's Tongue* (Reeks; rare).

465. *Echinospermum Lappula*, Lehm. (Reeks).

466. *E. Virginicum*, Lehm. (Reeks; rare).

467. *Mertensia maritima*, Don. *Sea Lungwort.* (Ice Plant, Labrador). (Reeks). Appears to be fairly common along most parts of the coast. I have found it in all, or nearly all, the open sea beaches where I have been. *Lab :* Hopedale (Weiz), and Caribou Island (Butler); several places in the Strait, at Indian Harbour and S. Michael's. *Flora Miq.*, very common. July, August.

468. *Myosotis laxa*, Lehm. Harbour Grace (Cat. II., 340); New Harbour, Harbour Breton, Exploits, and several places in the Bay of Islands (A. C. W.—Macoun and Fowler); Manuel's (R. & S.). Gardens and waste places. July, August.

469. *M. arvensis*, Hoffm. *Field Scorpion Grass or Forget-me-not.* S. John's (R. & S.), appearing as if introduced. *Lab :* Sandwich Bay (Revd. W. Shears—Macoun). A doubtful plant. August.

470. *M. palustris*, With. *Marsh or Great Forget-me-not.* Harbour Grace (Miss Trapnell—A. H. MacKay).

471. *Symphytum officinale*, L. *Common Comfrey.* Harbour Grace (McGill Coll. Herb., Cat. II., 343); S. John's (R. & S.).

LXVIII.—CONVOLVULACEÆ. *Bindweed Family* *

472. *Convolvulus sepium,* L. *Hedge or Great Bindweed.*
West Bay, Cape S. George (Bell), and Stevenville, Bay S. George
(A. White—A. H. MacKay); Spreadeagle in Trinity Bay, and
Topsail, Conception Bay (A. C. W.) July, August.

LXIX.—SOLANACEÆ. *Potatoe Family.*

473. *Solanum Dulcamara,* L. *Bittersweet.* S. John's (R.
& S.). Topsail road (Prof. Holloway—Fletcher).

LXX.—SCROPHULARIACE. *Figwort Family.*

474. *Bartsia alpina,* L. Lab: (Coln aster); Nachvak
(Bell, Cat. II., 367 ; III., 572).

475. *Chelone glabra,* L. *Snake or Turtle Head.* West of
Random, Trinity Bay (Cormack), and New Harbour in the same
Bay ; Frenchman's Cove, Bay of Islands (A. C. W.) ; Whitbourne
and Exploits River (R. & S.) ; S. John's (Miss Trapnell--MacKay) ;
The Goulds, near S. John's (Miss Southcott) ; (Cat. II., 354).
Wet places. August. *Flora Miq.*

476. *Castilleia pallida,* Kunth. *Painted Cup.* Lab:
Forteau (A. C. W.—Eaton). August.

Var. septentrionalis, Gray. In Bay of Islands at Chimney
Cove (Fowler), and Grand Lake (Robinson) by myself ; in island
at north side of Deer Lake (Bell) ; wet places and hills. July
—August. Lab: Hopedale (Weiz); Ford's Harbour and Nachvak (Bell—Packard).

477. *C. acuminata,* (Pursh) Spreng. Shoal Point (A. C. W.
—Coville). In grassy places. July. Dr. Robinson thinks that
this is the last-named plant.

478. *Euphrasia officinalis,* L. *Common Eyebright.* Appears
to be common in grassy places in many districts (Cat. III., 367).
Lab: (Cat. II., 367); Hopedale (Weiz-Packard) ; L'anse au Clair,

Battle Harbour and Fox Harbour (A. C. W.). *Flora Miq.*, very common. July, August.

Var. *Tartarica*, Berith. *Lab:* (Pursh), Caribou Island (Butler—Packard).

479. *E. purpurea (E. gracilis, Fries),* "new species." Sea coast at Cow Head. Much smaller in all parts; dark purple flower. (Reeks).

480. *Linaria vulgaris*, Mill. *Butter and Eggs.* Harbour Grace (Miss Trapnell). S. John's (Southcott and R. & S.); Harbour Breton (A. C. W.); La Scie, Notre Dame Bay, Revd. A. Pitman—Fowler). August.

481. *L. striata*, D. C. S. John's, on Rennie's River, but near waste heaps; doubtless a waif (R. & S.).

482. *Pedicularis palustris*, L. *Marsh Lousewort.* Moist meadows, S. John's (R. & S.). "The typical form of this does not appear to have been heretofore recorded in America. It differs from the var. *Wlassoviana*, Bunge, conspicuously in the form of the Corolla, and has also been collected in Labrador by Mr. J. A. Allen."

Var. *Wlassoviana*, Bunge (Cat. II., 369—Morrison); S. John's (Miss Southcott). *Lab:* Holton (A. C. W.) Bogs. July.

Var ? S. John's (A. C. W.—Mr. Howley and Macoun).

483. *P. flammea,* L. Ford's Harbour and Cape Chidley and Nachvak (Bell); Hopedale (Weiz, Cat. III., 573, and Packard).

484. *P. hirsuta,* L. *Lab:* Ford's Harbour and Cape Chidley (Bell, Cat. III., 376).

485. *P. Lapponica,* L. *Lab:* (Kolmeister, Cat. II., 368); Nachvak (Bell) and Hopedale (Weiz—Packard).

486. *P. pedicellata*, Bunge. *Lab:* (Gray, Cat. II., 368).

487. *P. euphrasioides*, Stephen. *Lab:* (Kolmeister, Cat. II., 368); Hopedale (Weiz), and Ford's Harbour (Bell—Packard); Sandwich Bay (Revd. T. Quinton); Holton (Revd. Wm. How—Macoun).

488. *P. Grœnlandica*, Retz. *Lab :* (Morrison, Cat. II., 368) ; Nachvak (Bell), and Hopedale (Weiz-Packard).

489. *Rhinanthus Crista-Galli*, L. *Yellow Rattle* (SHEP-HERD'S COFFIN). Appears to be very common throughout the country in grassy and wet places. *Lab :* abundant and very common in places on Bonne Espérance, and found all along the Labrador coast (Stearns, Cat. II., 371). July, August.

490. *Veronica Anagallis*, L. *Speedwell.* (Reeks). Hawk's Bay, N. W. Coast (Bullman). June, July.

491. *V. Americana*, Schwein. *American Brooklime.* Great Cod Roy River (Bell) ; in Bay of Islands at Riverhead (Fowler) ; Chimney Cove (Robinson) ; McIver's Cove (Trelease), collected by myself. Wet places. July, August.

492. *V. scutellata*, L. *Skull-cap Brooklime, Marsh Speed-well.* S John's (Lady Blake) ; muddy bank, Whitbourne (R. & S.) ; at Dildo, Trinity Bay (Macoun), and at Deer Lake near Bay of Islands (A. C. W—Fowler) ; (Reeks). Wet places. August.

493. *V. Buxbaumii*, Tenore. S. John's (Miss Southcott) ; Harbour Breton in Fortune Bay (A. C. W.).

494. *V. serpyllifolia*, L. *Thyme-leaved Speedwell.* (Reeks) ; New Harbour, etc., Trinity Bay, Topsail and Exploit (A. C. W.) ; S. John's (Lady Blake) ; Great Cod Roy River (Bell) ; Birchy Cove, Bay of Islands (A. C. W.—Fowler). *Lab :* L'anse au Clair (A. C. W.). Grassy and wet places. June and July.

495. *V. officinalis*, L. *Common Speedwell.* S. John's (Miss Southcott and R. & S.). July.

496. *V. alpina*, L. *Lab :* (Gray, Cat. II., 361) ; Port Bur-well, Cape Chidley (Bell, Cat. III., 571) ; Nain (Lundbery), and Hopedale (Weiz—Packard).

497. *V. arvensis*, L. *Corn Speedwell.* New Harbour (A. C. W.).

498.   *V. agrestis*, L.   Rocky hills, S. John's (R. & S.) : gardens, Birchy Cove, B. of I. (A. C. W.—Fowler).   August.

499.   *Mimulus luteus*, L.   *V. sessilifolius.*   Birchy Cove, B. of I. (A. C. W.—Fowler).   Brooklets.   August.   Dr. Robinson says this is *M. moschatus*, Dougl.

500.   *Scrophularia Marylandica*, (L.) Fries.   *Figwort.*   Meadows, B. of I. (A. C. W.—Coville).   Wet places.   July.

501.   *Pentstemon pubescens*, Solander.   *Beardstongue.*   Cow Head, W. coast (Bullman).   Dry soil.   June

502.   *Limosella aquatica*, L.   *V. tenuifolia*, Hoffm.   " *Meedwort*," Sterile, and accordingly doubtful specimens collected upon precipitous cliffs of Placentia Harbour (R. & S.).

LXVI.—HYDROPHYLLACEÆ.   *Waterleaf Family.*

503.   *Hydrophyllum Virginicum*, L.   Bonne Bay (Bullman).   Wet places.   July.

LXXI.—OROBANCHACEÆ.   *Broomrape Family.*

504.   *Aphyllon uniflorum*, Gray.   *Naked or One-flowered Broomrape.*   (Reeks).   (Miss Brenton, Cat. II., 372) ; Dildo (Rev. H. Petley—Macoun) ; Kilbride near S. John's (Miss Trapnell—A. H. MacKay) ; Coal River, B. of I. (A. C. W.—Fowler) ; Open woods.   July.

505.   *Conopholis Americana*, Walb.   *Squawroot.* (Reeks).

LXXII.—LENTIBULARIACEÆ.   *Bladderwort Family.*

506.   *Utricularia vulgaris*, L.   *Common Bladderwort.*   Whitbourne and Exploits River (R. & S.).

Var. *Americana*, Gray (Reeks) ; West of Random (Cormack); also in Trinity Bay, at New Harbour (A. C. W.).   Ditches and brooks.

507.   *U. minor*, L.   (Reeks ; rare).

508. *U. intermedia*, Hayne. Brook, Placentia (R. & S.); about New Harbour and Broad Cove, Trinity Bay, Harbour Breton, Western Cove, White Bay, Great Harbour, Hermitage Bay (A. C. W.). July—August. *Flora Miq.*, stagnant waters, Common.

509. *U. cornuta*, Mx. *Horned Bladderwort*. Sphagnous or sandy swamps from Newfoundland to Lake Superior (Gray, Cat. II., 376); (Reeks) rare : S. John's (Lady Blake); Chimney Cove and Grand Lake, and other places in Bay of Islands ; New Harbour and Harbour Breton (A. C. W.); Whitbourne and Exploits River (R. & S.). *Flora Miq.*, common. July, August.

510. *Pinguicula, villosa*, L. *Lab :* (Gray, Cat. II., 376); Hopedale (Weiz-Packard).

511. *P. alpina*, L. *Lab :* Steinhauer; not elsewhere detailed in America (Gray, Cat. II., 376); Hopedale (Weiz-Packard).

512. *P. vulgaris*, L. *Common Butterwort*. (Miss Brenton, Cat. II., 376); Burin graveyard, and not uncommon about Bay of Islands (A. C. W.); Flat Bay Brook, Bay S. George (Bell). *Lab :* (Cat. II., 376); L'anse Amour Bay (Butler); Hopedale (Weiz); Nachvak (Bell—Packard); Forteau, Battle Harbour, Seal Islands, Snack Cove, Holton (A. C. W.). Wet rocks. June —August. *Flora Miq.*, common.

LXXVII.—LABIATÆ. *Mint Family.*

513. *Brunella vulgaris*, L. *Seal Head*. (Reeks). Great Cod Roy River (Bell) ; Trinity Bay and several places about Bay of Islands and at Harbour Breton (A. C. W.), near Salmonier River; common ; (R. & S.) ; S. John's (Miss Southcott). *Flora Miq.*, very common. July, August.

514. *Collinsonia Canadensis*, L. *Horseweed*. (Bonycastle).

515. *Calamintha clinopodium*, Benth. *Wild Basil*. Rich bottoms, Salmonier (R. & S.) ; Chimney Cove, B. of I. (A. C. W. —Robinson). Grassy hills. August.

516. *Galeopis Tetrahit*, L. *Dead Hemp Nettle.* S. John's (Miss Southcott and R. & S.) ; (Reeks) ; New Harbour and elsewhere in Trinity Bay (A. C. W.—Cormack) ; Chimney Cove, Birchy Cove and Middle Arm, Bay of Islands (A. C. W.—Fowler); Riverhead, White Bay (Bullman). Roadside and gardens. July, August.

517. *Galeopsis Ladanum*, L. *Hemp Nettle.* (Reeks) ; S. John's (Miss Southcott).

518. *G. versicolor*, L. S. John's (Miss Southcott).

519. *Lycopus Virginicus*, L. *Bugle weed, Virginian Horehound.* West of Random and in Trinity Bay (Cormack : and New Harbour (A. C. W.) ; S. John's (Miss Southcott), and rocky banks, Rennie's River (R. & S.) ; Sandy Point, Bay S. George (A. C. W.—Fowler), and a few places in Bay of Islands. Wet places. August. *Flora Miq.*, common.

520. *L. sinuatus*, Ell. Salmonier River (R. & S.) Gravel beds. August.

521. *Lamium amplexicaule*, L. *Henbit Dead Nettle.* New Harbour (A. C. W.) ; S. John's, fields (R. & S.). August.

522. *L. purpureum*, L. *Red Dead Nettle.* S. John's (Miss Southcott) ; New Harbour and Harbour Breton (A. C. W.).

523. *L. maculatum*, L. New Harbour (A. C. W.) Gardens.

524. *L. incisum*, Willd. S. John's ; a single specimen by roadside (R & S).

525. *Mentha viridis*, L. *Spearmint* (Reeks).

526. *M. arvensis*, L. *Cornmint.* Manuel's River, rocky banks ; common along streams (R. & S.) August.

527. *M. Canadensis*, L. *Canada or Horse Mint.* New Harbour, Trinity Bay (A. C. W.), and elsewhere in the same Bay (Cormack) ; Flat Bay (Bell) ; (Reeks) ; Chimney Cove, and Irishtown, B. of I. (A. C. W.—Fowler). Wet places. August.

Var. glabrata, Benth. Spreadeagle, Trinity Bay (A. C. W.).

528. *Nepeta Cataria*, L. *Catnip, Catsmint.* John's Beach, B. of I. (A. C. W.). Roadside. August.

529. *N. Glechoma*, Benth. *Ground Ivy, Gill over-the-ground* (SCARLET-RUNNER). Harbour Grace (McGill Coll. Herb., Cat. II., 387); S. John's R. & S.): Topsail, old shop (Trinity Bay), and Bay de Verde (A. C. W.) *Lab*: Battle Harbour (A. C. W.). Roadside and old gardens. June—August.

530. *Scutellaria lateriflora*, L. *Maddog Scullcap*. (Reeks). Deer Lake, near Bay of Islands (A. C. W.—Fowler). River banks. August.

531. *S. galericulata*, L. *Common Skullcap*. (RED TOPS). (Cat. II., 388). Trinity Bay (Cormack); Green Harbour and other places in the same Bay (A. C. W.); (Reeks); Manuel's River, rocky banks (R. & S.); Salt Water Pond, White Bay (Bullman). Sea beach. July—August.

532. *Stachys palustris*, L. *Woundwort* (Cormack); Harbour Grace (Miss Trapnell—A. H. MacKay); S. John's (R. & S., wet meadows); Sandy Point, Bay S. George, gardens (A. C. W. —Fowler. Wet ground, from Newfoundland to the Pacific (Gray, Cat. II., 390. August.

N. B.—The *Flora Miquelonensis* says that Thymus vulgaris, L., Satureia hortense, L., Galeopsis Ladanum, L., Mentha piperita, L., Lamium amplexicaule, L., have been introduced and are found in gardens or in the neighbourhood.

LXXVIII.—PLANTAGINACEÆ. *Plantain Family.*

533. *Plantago major*, L. *Common Plantain* (RAT-TAIL.) (Reeks); Great Cod Roy River (Bell); S. John's (R. & S.); Birchy Cove, B. of I. (A. C. W.). *Lab*: Battle Harbour (A. C. W.) Roadsides. August.

534. *P. eriopoda*, Torr? Prof. Macoun thinks this may be Dr. Bell's P. Virginica.

535. *P. maritima*, L. *Seaside Plantain.* Trinity Bay (Cormack); New Harbour and other places about Trinity Bay, and at Middle Arm, and in Bay of Islands (A. C. W.); Placentia

(R. & S.).   Dr. Bells says a large variety (?) was found with broad leaves and long tapering point near extremity of Flat Bay. *Lab:* (Pursh, Cat. II., 393) ; Nachvak (Bell, Cat. III., 575) ; Hopedale (Weiz) ; and Caribou Islands (Butler-Packard).   Sea cliffs.   July, August.   *Flora Miq.*, very common.

536.   *P. decipiens*, Barneoud.   Port a Port (Bell) ; crevices of rocks.   (P. maritima, var. juncoides, Gray).   *Lab:* (Gray, Cat. II., 393).

537.   *P. lanceolata*, L.   *Ribwort Plantain.*   (Reeks) ; Middle Arm and Birchy Cove, B. of I. (A. C. W.—Trelease and Robinson) ; S. John's (R. & S.).   Fields and gardens.   June—August.

538.   *Littorella lacustris*, L.   *Plantain Shoreweed.*   Exploits River (R. & S.).   Muddy banks.   August.

The *Flora Miq.* remarks that *P. major*, L. and *P. lanceolata*, L., are common around houses ; probably introduced.

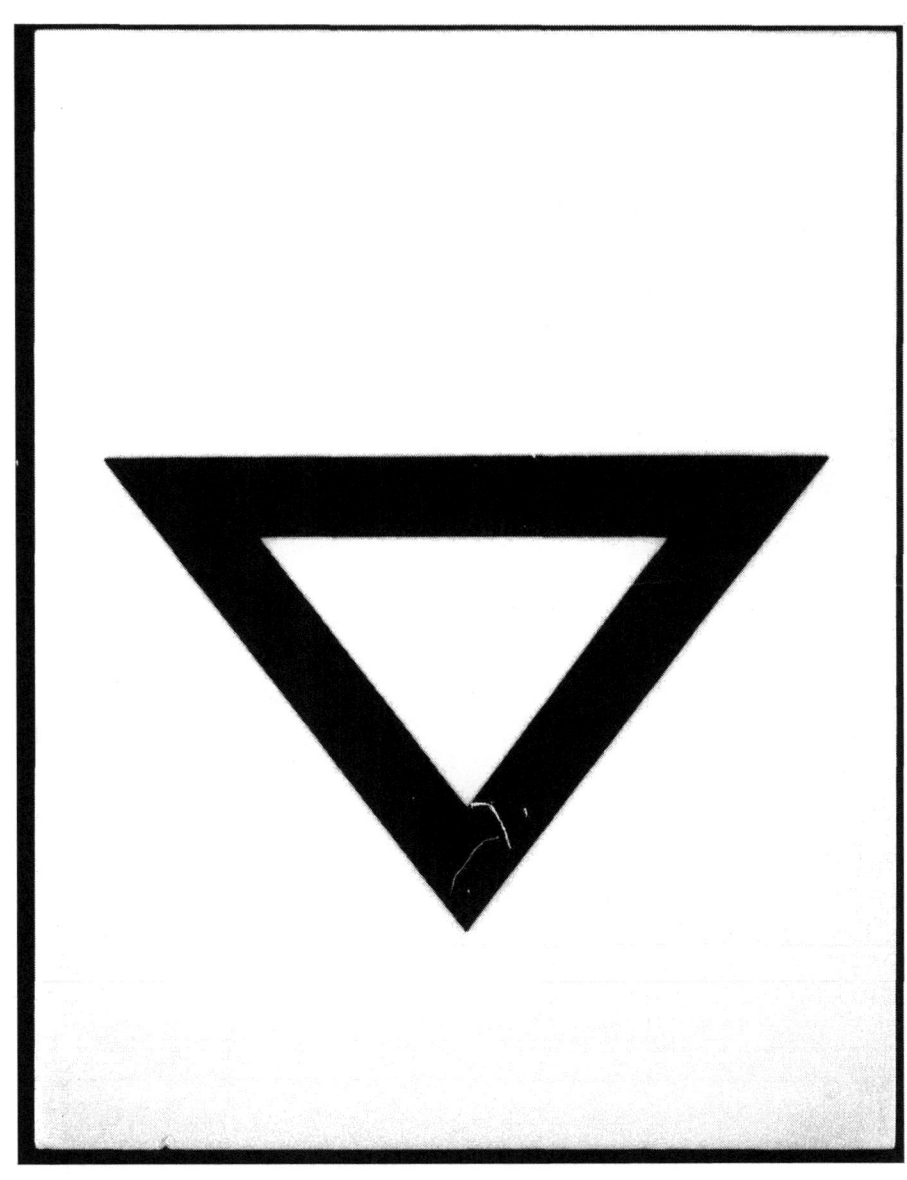